YOUR KNOWLEDGE HAS VALUE

- We will publish your bachelor's and
  master's thesis, essays and papers

- Your own eBook and book -
  sold worldwide in all relevant shops

- Earn money with each sale

Upload your text at www.GRIN.com
and publish for free

Bibliographic information published by the German National Library:

The German National Library lists this publication in the National Bibliography;
detailed bibliographic data are available on the Internet at http://dnb.dnb.de .

Imprint:

Copyright © 2010 GRIN Verlag
Print and binding: Books on Demand GmbH, Norderstedt Germany
ISBN: 9783346068422

This book at GRIN:

https://www.grin.com/document/506250

Albert Johan

Creating microstructures using conducting polypyrrole

Voltammetry method by varying voltage, cycle times, surfaces

GRIN Verlag

GRIN - Your knowledge has value

Since its foundation in 1998, GRIN has specialized in publishing academic texts by students, college teachers and other academics as e-book and printed book. The website www.grin.com is an ideal platform for presenting term papers, final papers, scientific essays, dissertations and specialist books.

Visit us on the internet:

http://www.grin.com/

http://www.facebook.com/grincom

http://www.twitter.com/grin_com

Table of Contents

1 – Abstract

Simple electrochemical aided polymerization was done using pyrrole and beta-napthalenesulfonic acid as electrolyte *cum* anionic surfactant to attempt creating microstructures on sputtered gold surface, highly oriented pyrolitic graphite, and thiol-modified gold surface. Hydrogen bubble template was used on gold, but results showed doubted polymerization around this. Instead, one tenth structure size of hydrogen bubbles was commonly observed in two different electrolyte concentrations. No interesting microstructure was formed on highly oriented pyrolitic graphite. This was due to polarization of large size graphite. Irregular microcapsules were formed around n-decane template on thiol-coated gold, but other structures also appeared. Lack of control on many variables due to limitation of equipments and materials lead to inaccuracy and inability to do more detailed observation. However, some results showed research aim was achieved.

2 – Introduction

Since last decades, extensive research on fabrication of microstructures of conducting polymers has been aimed to exploit their semiconductive, optical, flexibility, biocompatibility, nano/microscale, low density, and large surface area properties[49]. The term conducting polymers refers to polymers with conjugated chain structure with semiconductive property due to alternating single and double bond[54]. Microstructures differ with their size, shape, and applications. Also, chemical and electrochemical pathways and various templating methods offer their own unique advantages and disadvantages in forming desired microstructures[1]. Obviously, narrowing down the scope of this research is a sensible step to focus investigating fabrication of microstructures of conducting polymer.

One feasible research proposal was to investigate gaps in what have been found on microcontainers fabricated by simple electrochemical method using polypyrrole as conducting polymer. Microcontainers could be defined as polymerized shape that could be further encapsulated and reopened (e.g cups, bowls, bottles, capsule, spheres), mostly, reversibly[1]. Many studies have been able to control morphology and properties of microcontainers by modifying electrochemical condition, applied potential, concentration, and techniques (for example, solidified droplet, etc). Conflicting and inconsistent evidence have been seen, for example, on applied potential to form soap bubble and start polymerization. Most importantly, gaps appeared in no formation of microcontainers on some electrodes: gold and highly orientated pyrolitic graphite (HOPG). Also, there has not been any electrochemical polymerization around oil-droplet template (n-decane is used in this project).

This research project aims to (1) electropolymerise pyrrole around hydrogen bubble template formed on gold surface (2) electropolymerise pyrrole around n-decane template on highly oriented pyrolitic graphite and thiol-coated gold surface, and (3) generally link some conditions of polymerization to microcontainer shapes formed. If sputtered gold was stable under polymerization potential, this surface would produce products similar to that of stainless steel surface. If working electrode size should be equal or less than counter electrode to avoid polarization[1], then highly oriented pyrolitic graphite would possibly be polarized because it was significantly bigger than platinum counter electrode. And if anionic surfactant beta-naphthalenesulfonic acid coated n-decane droplet, polypyrrole backbone would attach around n-decane template during electropolymerisation.

3 – Literature Review

Narrowed down by the project scope, this review only included literatures promoting microstructure formation using simple equipments, materials, and methods. One example is electrochemical polymerisation of pyrrole around soft-templates in beta naphthalenesulfonic acid as electrolyte with various working electrodes and conditions. Whereas, more advanced equipments, materials, and methods were outside this project scope and hence, excluded. This review would consider all aspects of a simple electrochemical process such as available equipments and materials, electrochemical conditions, electrode material, hydrophobicity, and size. Conflicting evidence was noted in applied potential to both produce soap bubble templates and start polymerisation, and gold hydrophobicity as working electrode. Gaps appear in no formation of microcontainers on gold and highly oriented pyrolitic graphite, and no electrochemical polymerisation around oil-droplet template.

Materials and equipments were available to do simple controlled electrochemistry. Potentiostat and three electrode cell (working, counter, and reference electrode), pyrrole, beta-naphtalenesulfonic powder, OTS-silicon, gold and chromium sputter, highly oriented pyrolitic graphic, and 1mM thiol solution were main basic materials and equipments available to control variables to determine product properties. Stainless steel as common working and counter electrode[1] was not available. As counter electrode, platinum wire-coil could be used as common substitute to stainless steel[1]. Gold surface was used as substitute for working electrode because it is inert to electrolyte has good conductivity[1]. Polymerisation could be done using chemical or electrochemical paths[1]. Electrochemical path was chosen because it offers greater control over variables to produce desired microstructures and could be done in ambient environment[1]. Using available potentiostat, electrochemical polymerisation reaction rate can be controlled by the applied potential or current density. Additionally, the amount of product can be controlled by the integrated charges used for electrosynthesis. Furthermore, product's morphology and property can be modified using electropolymerisation conditions such as electrolyte and pyrrole concentration[1], and different templating methods [55,56,1].

Usage of pyrrole as monomer and beta-naphtalenesulfonic acid as electrolyte *cum* surfactant, hydrogen bubbles, and n-decane as soft-template was feasible for this project. Pyrrole was chosen due to its well known conductivity, chemical stability, mechanical properties[57], and most importantly, availability[4,5]. Template shape, to which pyrrole backbone will attach to during polymerisation, would also determine microstructure shape as product[1,5,4,49,54-56]. Beta-naphthalenesulfonic acid is anionic surfactant and electrolyte at the same time. This was known to coat hydrogen bubble with negative charges, and allow deposition of bubbles on working electrode when positive potential was applied at the working electrode. At this same instant, pyrrole underwent redox reaction to become conductive polypyrrole which acquired anion by attaching its backbone to surfactant-coated bubble. Thus, polymerization occurred around hydrogen bubble template[1]. The available microscope limits clear observation of product's morphology only up to 10 micrometer scale. Various template methods with size≥10 micrometer that have been recognised were layer by layer[48], oil droplet [56], solidified droplet[55], microbubbles [1,2,3], membrane and solid[1], nanosheets[1], and micelles[45,46]. This list was further reduced by selecting the ones offering easiness to form before/during and remove after each experiment. Finally, simplest soft-templates such as hydrogen bubbles (or soap bubbles) and n-decane were chosen due to considerable practicality they offer. Easily, soap bubbles could be produced during experiment by water electrolysis and n-decane could be prepared by direct surface adsorption in 20%v/v ethanol emulsion[4]. Hydrogen bubbles and n-decane templates could be removed after each experiment by, respectively, exposure[1] and drying with atmospheric air[4]. Microstructures formed by corresponding soft-templates were known as "microcontainers" with potential application of sensing, attributed to their large conductive surface area[2], and reversible encapsulating and releasing organic species such as dyes and drugs[1].

There has been abundant studies linking electrochemical condition to produce gas bubbles and start polymerisation of pyrrole with microcontainer morphology. Most studies were in agreement with others and some were contradictive and inconsistent. Firstly, increasing either applied potential or pyrrole concentration would increase "deformation force". This deformed microcontainer shape from spherical to elliptical and then to cylindrical[49]. Additionally, extensive proofs on increased density, shape, and wall thickness of polypyrrole were observed by, respectively, increasing either the width of applied potentials for generation of hydrogen bubbles or concentration of electrolyte, increasing either scanning rate, width of applied potentials to start polymerisation, or concentration of electrolyte, and increasing either cyclic voltametric scan number, concentration of pyrrole, or scanning rate[2]. Contradictions appeared on effective applied potential to produce soap bubbles as template. Cyclic voltametry was commonly used because One study used cyclic voltametry from 0V to 1.2V for various cycles with onset oxidation potential of water of -1.25V (versus saturated calomel electrode) to form soap bubble template and microcontainers

5

simultaneously[3]. This method was judged to have a detrimental overoxidation effect because it directly generated oxygen gas on the working electrode under a relatively high positive potential (>0.8V)[47,2]. The detrimental effect was predicted to be polypyrrole film[4,48] or droplet[48] deposition on working electrode. Pyrrole droplet was another templating method that produce microcontainer of around 10 micrometer size[48], but considered as a different method. Applied negative potential of −1.0 to −1.4 V[47] and -1.0V to -1.3 or -1.4 or -1.65 or -1.7V[2] for first cycle were proven effective to generate hydrogen bubbles as template. Furthermore, there has been inconsistency in applied potential for polymerisation of pyrrole. One study claimed that no microstructures were formed under 0.8V applied potential[3]. This is inconsistent to a finding that successfully formed microcontainers by applying -1.0V to -1.7V for first cycle, continued with 0.5V-0.75V for the next ten cycles using 100mV/s cyclic voltametry[2]. The reason of this inconsistency seemed to be former study's weakness by not producing any soap bubble template under 0.8V potential because of the absence of first scan of negative potential.

Stainless steel[2,3,47] and platinum[49] were common working electrodes in production of soap bubble-templated microcontainers. However, there has not been any (or, only a few, if existed) evidence of microcontainers produced on other electrode material such as gold[1] and highly oriented pyrolitic graphite[25]. Instead of microcontainers, evidence has shown film polymerisation on highly oriented pyrolitic graphite[61] and gold[60]. One possible explanation seemed to be these material was not suitable in electropolymerisation systems, purpose of experiments, surface properties, shapes, and costs[1] to produce microcontainers. Nevertheless, it would be interesting to investigate whether microcontainers could actually be formed on gold and highly oriented pyrolitic graphite electrodes. This research aimed to form gas bubble-templated microcontainers on gold surface, but not on highly oriented pyrolitic graphite. The decision not to use soap-bubble template on highly oriented pyrolitic graphite would be discussed in later paragraphs.

As another soft-template than gas bubbles, oil droplet has received little attention. One reason could be the new method of solidified droplet seemed to be more promising than liquid droplet [and also possibly soap bubble] because it offered stronger microstructure to avoid wall collapse and inhibited degradation of dye/drugs while sealing the mouth of microcontainers[55]. Nevertheless, both oil and solidified droplet templating lied on the same principle of surface adsorption, which created water-insoluable template on hydrophobic surface[55,56]. Surface hydrophobicity played important role in adsorbing non-polar liquid[4,5], for example, n-decane[4], mineral oil[56], and heated tetradecanol[55]. Two studies successfully adsorped dye and heated tetradecanol[55]/mineral oil[56] on quartz glass, as well as sealing them in hemispherical polypyrrole capsule by chemical polymerisation path[55,56]. There were two interesting questions that arised from this finding. Firstly, what caused the hemispherical shape? One claimed this was due to hydrophobic glass interaction that affect non-polar oil template shape[4]. If this was true, one could observe oil-droplet shape on hydrophobic surface and expect this to be template shape for polymerisation. This research measured oil-droplet contact angle in atmosphere and check if this could be related to template shape. Secondly, would electrochemical polymerisation path also work? This research aimed to electrochemically polymerise pyrrole around n-decane template. Only positive cyclic voltametric scan was required because gas bubble production template was replaced by oil. n-decane was used because it fulfills the requirement of non-polarity and immiscibility in water[4]. One unknown factor to be observed was whether anionic surfactant beta-naphthalenesulfonic would coat n-decane bubbles during direct surface adsorption. This would affect pyrrole backbone attachment, and hence, structure created. Considered surfaces are hydrophobic surfaces so n-decane could adsorb on it[55,56].

To do this experiment, while stainless steel would have been ideal as working electrode because its surface was well known to be highly hydrophobic (contact angle with water = $\theta_{H2O} \sim$ 79°)[59], gold hydrophobicity was found to be highly debated. Since 1934, 26 (or more) attempts have been made to establish the hydrophilic or hydrophobic nature of clean gold surfaces. Eighteen (including theoretical studies) concluded clean gold is hydrophobic ($\theta_{H2O} \sim$ 65°), whereas eight concluded clean gold is hydrophilic[58]. Other studies showed that bare gold was hydrophobic with relatively large reported contact angle within the range of 40 to 79.5 degrees[16] and 65.5 degrees[19], which was unexpectedly high for metal[30]. Conclusive study explained this phenomena: less than a monolayer of carbonaceous contamination renders the gold surface hydrophobic ($\theta_{H2O} \sim$ 50°), whereas clean gold is hydrophilic[58]. Because gold carbonaceous contamination is an uncontrolled factor during experiment, it was safer to modify gold surface. Firstly, it is assumed that bare gold was hydrophilic in water. Afterwards, thiol-coating was applied to create covalent bond between thiol and gold. Self assembly monolayer (SAM) of thiol made gold surface hydrophobic. This method could guarantee successful n-decane direct surface adsorption on hydrophobic surface to create oil droplet[4].

Highly oriented pyrolitic graphite was reported hydrophobic[26] ($\theta_{H2O} \sim 90°$[4]), therefore it seemed suitable to use this as working electrode for n-droplet template. Furthermore, due to its limited availability and high cost, it was not used to polymerise hydrogen template. The size of highly oriented pyrolitic graphite was bigger than platinum coil as counter electrode. Additionally, its effective surface areas were both sides of the electrode. One study implied this would cause polarisation (or often referred as overpotential) of working electrode because it is larger than the counter electrode[1]. Polarisation was defined as production of thermodynamically irreversible potential on electrode surface, which speeds up reaction and might cause overoxidation[31]. In other words, an expected effect due to larger size of working electrode relative to counter electrode is overpolymerisation of pyrrole on highly oriented pyrolitic graphite electrode. This occurence would seem to cause difficulty in observation of microstructures. This implication has been discussed before conducting experiment but this study was not taken into consideration[4].

In conclusion, many studies have contributed to control morphology and properties of microcontainers by modifying electrochemical condition, applied potential, concentration, and formation techniques. The knowledge has been applied in many fields such as medicine, communication, etc. In scope of simple electrochemical approach, gaps were observed in no microcontainer formation on some common electrodes such as gold and highly oriented pyrolitic graphite. Also, electrochemical pathway to polymerise around oil-template seemed absent. This review had attempted to fill knowledge gap in understanding simple electrochemical approach of fabricating microstructures.

4 – Experimental section

4.1 – Methods

Surface cleaning: glass vial and petri dish to store electrode and solution
10% (g/mL) NaOH solution was used to react with glass outer layer and remove it. This solution is poured to fulfill vials and glass petri dish and kept at 30 degree Celcius for 30 minutes. Afterwards, glass is rinsed in Mill-Q water and dried in oven using (e.g 100 degree Celcius in 5 minutes)[4].

Bare gold preparation

Cutting

Bare gold was cut using diamond cutter to smaller pieces than platinum counter electrode. This is so that polarisation of working electrode is reduced[1]. Another reason is so it fits the hole to insert electrodes in electrochemical cell. Stainless steel tweezer was used to move and hold the working electrode with care.

Cleaning
Bare gold was rinsed with ethanol and water, then dried using nitrogen gas.

Gold attachment to copper wire
One side of bare gold was touched to copper wire. They are wrapped together using teflon tape.

Thiol-coating on gold
Bare gold was cleaned using ethanol and water. Afterwards, it was put under UV light for further 20 minutes to remove organic layer. This was immersed in 1mM of thiol solution for 1 hour so thiol formed covalent bond with gold (chemisorption). This modify gold surface hydrophobic. Afterwards, thiol-coated gold was left dry and attached to copper wire[4].

Working electrode Octadecyltrichlorosilane (OTS)-Silicon surface preparation
1. Surface cleaning: silicon surface
Piranha solution (7:3 v:v H_2SO_4 98%: H_2O_2 30%) was put on approx. 5 x 1.5 cm^2 silicon surface. This solution is a strong oxidiser and was prepared with extreme caution. The

temperature of solution and surface was kept at 75 degree celcius for 20 minutes. Afterwards, silicon is rinsed with and immersed in Mill-Q water. This treatment will let piranha solution to react with all organic substance in Mill-Q water, including covalent bonded self assembly monolayer (SAM) on the surface. This method does not roughen the surface[4].

2. Surface sputtering

Gold surface was formed by coating OTS Silicon with chromium and gold using available sputter. A middle thin chromium layer was necessary to coat OTS-Silicon because gold does not bind well with OTS-Si[4]. 40-50nm Au and 5nm Cr coating on OTS-Silicon surface was prepared using EMITECH K575 turbo sputter. The steps are:

1. Setting up current 60mA and time 30s for chromium coating on OTS-Si
2. Vacuum was created on chamber area
3. 5nm layer of chromium coated OTS-Si
4. Atmospheric pressure was restored, setting up current 70mA and time 120s for gold coating on Cr-coated OTS-Si.
5. Vacuum was created on chamber area
6. 40-50nm layer of Au coated Cr-coated OTS-Si.
7. Atmospheric pressure was restored
8. Working electrode surface was produced and is referred as "bare gold" and left in cleaned petri dish for a good weekend before used.
9. Machine broke down sputtering due to overcurrent[4].

Working, reference, and counter electrodes size and position

The reference electrode is used to refer the potential of working electrode during polymerisation and should be placed close to working electrode [1]. Counter electrode size was relatively bigger than gold so that polarization on working electrode is reduced[1].

Direct surface adsorption of n-decane droplets on working electrode surface

20 microliter (approximately 2 small drops) of n-decane was taken using stainless steel syringe and mixed rigorously with 20%v/v ethanol 20mL solution in a vial to form decane bubbles in cloudy emulsion. Thiol-coated gold or highly oriented pyrolitic graphite as working electrode was lied horizontal on the bottom of electrochemical cell and immersed with this emulsion, allowing decane droplets to adsorb immediately on surface. Surface is positioned horizontally to allow decane droplets formation with similar right and left contact

angle. After one minute, most of emulsion is discarded from cell, leaving a thin emulsion layer (negligible volume) still immersing the surface[4]. Surface is positioned on the bottom to control constant distance between working and reference electrode as required[1]. Unfortunately, to do this, it is impossible to avoid immersion of copper wire in electrolyte. Microstructures would be formed too on copper wire during electrical polymerisation.

Position of working electrode
Gold was dipped vertically in electrolyte and positioned as close as possible to reference electrode. In this arrangement, copper wire attached to gold could avoid contact with electrolyte. However, as a trade-off, it is extremely difficult to fix working electrode distance to reference electrode.

n-decane-adsorped surface position

Surface with n-decane adsoped on it was positioned horizontally to allow decane droplets to have same right and left contact angle. After one minute, most of emulsion is discarded from cell, leaving a thin emulsion layer (negligible volume) still immersing the surface[4]. Surface is positioned on the bottom to control constant distance between working and reference electrode as required[1]. Unfortunately, to do this, it is impossible to avoid immersion of copper wire in electrolyte. Microstructures would be formed too on copper wire during electrical polymerisation.

5 minute degassing
Degassing was done to substituting of dissolved oxygen with inert gas, because this could disturb polymerization result. Nitrogen line was connected to degassing unit, which extended small teflon tubes of nitrogen gas to electrolyte and pyrrole aqueous solution.

Pyrrole distillation
10mL old pyrrole (99%v/v, Acor) was gently boiled in fume hood to 129 degree Celcius (boiling point). The distillant is cooled by ambient temperature of air and has no colour.

Mixing pyrrole and beta-naphthalenesulfonic acid
Pyrrole did not form aqueous solution with beta-napthalene sulfonic acid. If allowed to sit, they separated within 5 minutes with pyrrole at the top layer due to lower density. Therefore, they were mixed by shaking this solution rigorously in a clean vial before put in electrochemical cell. Experiment was done before they separated.

Sessile static contact angle measurement
Solution of interest was taken using microliter syringe and droplet as much as needle-tip was put on surface of interest. Immediately, right and left contact angle was measured using goniometer.

4.2 – Set-up

Measuring gold surface stability
Control set up: gold surface before polymerization
Experimental set up: gold surface after polymerization
Gold stability was measured by comparing gold surface before and after polymerization occurred. High stability means no change in gold surface[4]. Observation was done on gold surface, not on structures or polymer formed.

Measuring gold surface hydrophobicity
Control set up: thiol-coated gold surface
Experimental set up: bare gold surface
Thiol coated-gold is known hydrophobic and bare gold hydrophobicity was measured before used as electrode and compared to that of thiol-coated gold. If gold hydrophobicity is close to that of thiol-coated gold, gold was contaminated with carbonaceous layer.

4.3 – Condition

Electropolymerisation of pyrrole around hydrogen bubbles on bare gold surface
Condition 1
Beta-naphthalenesulfonic acid concentration = 0.2M
Condition 2
Beta-naphthalenesulfonic acid concentration = 0.4M

Pyrrole concentration = 0.25M

Cyclic voltametry scan of -1.0 to -1.7V for first cycle and 0.5V-0.9V for 2 cycles

First cycle to produce hydrogen bubbles, and subsequent two cycles to start polymerizing in low potential to avoid overoxidation.

Electropolymerisation of pyrrole around n-decane bubbles on highly oriented pyrolitic graphite

Condition 3

Cyclic voltametry scan of 0.5V-0.9V for 2 cycles

Electropolymerisation of pyrrole around n-decane bubbles on thiol-coated gold

Condition 3

Cyclic voltametry scan of 0.5V-0.9V for 2 cycles

Condition 4

Cyclic voltametry scan of 0.4V-0.6V for 2 cycles

Pyrrole concentration = 0.25M

Cyclic voltametry scan of -1.0 to -1.7V for first cycle and 0.5V-0.9V for 2 cycles

First cycle to produce hydrogen bubbles, and subsequent two cycles to start polymerizing in low potential to avoid overoxidation.

4.4 – Materials

Working electrode

1. Highly Oriented Pyrolitic Graphite (ZYH, NT-MDT, Moscow, Russia)
2. Silicon surface spurred with chromium and gold, some pieces further modified with thiol

All working electrode are attached to copper wire and tighten with teflon tape. With all means possible, copper wire was not immersed into electrolyte.

Counter electrode

Platinum wire welded to platinum coil was used as counter electrode. After most experiments conducted, welding was found loose and rewelded. This could contribute inaccuracy in current i(mA)reading at i(mA) vs E(V) curve

Reference electrode

Ag/AgCl with inner electrolyte made using saturated KCl solution in water. Saturation was done by adding excess KCl to water and sonication. During experiment, reference electrode was often blocked by solid depositation of electrolyte and polypyrrole. This made working electrode voltage E(V) reading sometimes inaccurate on i(mA) vs E(V) curve. Electrode was cleaned and solution was changed three times during experiments.

Pyrrole 99%v/v, Acor

Diluted to 0.25M, and twice distilled during experiment.

Electrolyte cum Anionic Surfactant

Beta-naphthalenesulfonic acid 70%w/w, dissolved in Milli-Q water to required concentration.

Water

1. 18.2 Mega ohm Milli-Q water mostly used to form solution such as dissolving beta-naphthalenesulfonic acid powder into required solution concentration, and n-decane emulsion in 20%v/v ethanol.
2. Distilled water was mostly used to wash electrode surface and clean vial/petri dish

4.5 – Schematic diagram

All electrochemical reaction is conducted at ambient temperature and pressure in an electrochemical system (Figure 1). Each electrodes are inserted into cell through small holes which are closed after insertion. When n-decane was not used as template, gold working electrode is carefully dipped to avoid copper wire immersion. When n-decane was used, working electrode was unavoidably immersed and placed horizontally at the bottom of cell. Refer to Method section 4.1 for explanation.

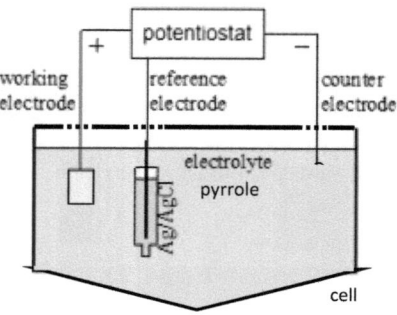

Figure 1 – simple electrochemical system

14

5 – Result and Discussion

5.1 – Gold surface stability

Gold surface before polymerization was considerably homogenous (Figure 2a). After polymerization, gold surface was found to change by having brown circles (Figure 2b).

2a

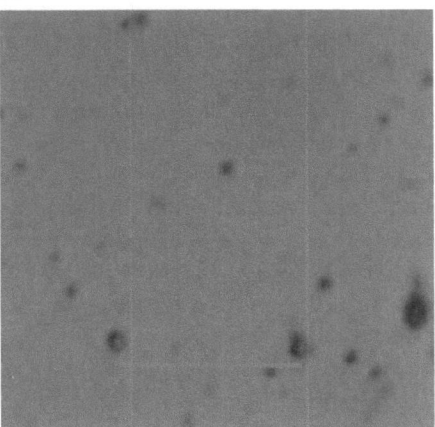

2b

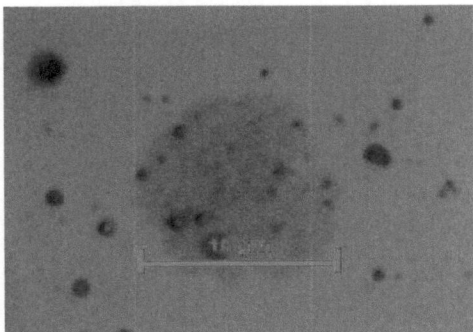

Figure 2 – a) Gold surface before polymerization, scale bar = 10 micrometer b) Brown circle on gold surface after polymerization (observed a few days later)

This brown circle was not polymer structures and was considered as change on surface property[4]. Because surface property changed, sputtered gold was unstable[4]. Instability could mean during polymerization, surface property was heterogenous and conducting uneven charge. This is not ideal condition as working electrode[1], especially, because it could affect many things such as reaction in electrode, bubble attachment, and uneven rate of polymerization, or even worse, dissolving gold layer.

5.2 – Electropolymerisation of pyrrole around hydrogen bubble template on bare gold surface

Microcups with approximate average size of 5-20 micrometers were successfully produced (Figure 3) on gold surface. The size was about one-tenth of common microcontainers produced on stainless steel surface using gas bubbles as template (~100 micrometers)[2]. Since it has been recognized that gas bubble sizes as template are around 100 micrometers[48,2], polymerisation was cleary not formed around gas bubbles. Instead, the structures could have been formed due to attachment of pyrrole droplet that polymerised on gold surface, either direct or with polyelectrolyte coating[48]. If this was true, eventhough the product size was in agreement(~10 micrometer), the density was still much less than predicted[48]. Also, there were different density of structures in different area. Lower pyrrole concentration (0.2M than 1.0M[48]) could be one explanation, but unstable electrode would possibly be another factor for this difference.

Increasing electrolyte concentration from 0.2M(Figure 3) to 0.4M(Figure 4) seemed to give predicted effect of increased shape (from microcup into microcontainers and microballs) and density[2]. Interesting ~100 micrometer structures were produced in Figure 4B, and the size were close to hydrogen gas. However, the shape was too irregular to conclude that this was caused by hydrogen bubble template[2]. The structure was interesting because it was dark in the middle, suggesting this could be a hole, or pyrrole droplets[48] forming a lump and polymerized on gold surface. Surface instability could also affect this shape formation. If more advanced imaging or characterisation equipment had been available, such as side atomic microscope, real time microscope to observe polymerization process, or raman spectroscopy, these structures could have been observed in more detail. Linear increase in negative potential lead to hyperbolic increase of negative current (Figure 5) until it reached ~(-1.5V,-0.9mA) and then it was out of scale range. This implied increase in conductivity during this first cyclic voltametry scan that might be attributed to anionic ion

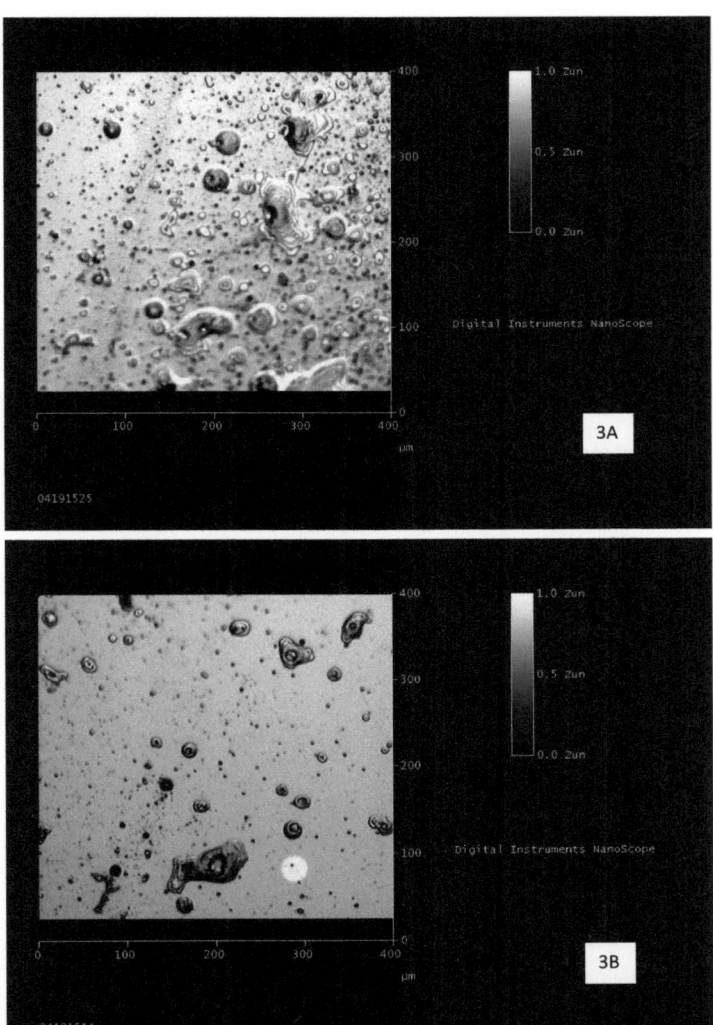

Figure 3 – Microcontainers formed by electropolymerisation of pyrrole on bare gold using condition 1: Beta-naphtalenesulfonic acid concentration = 0.2M

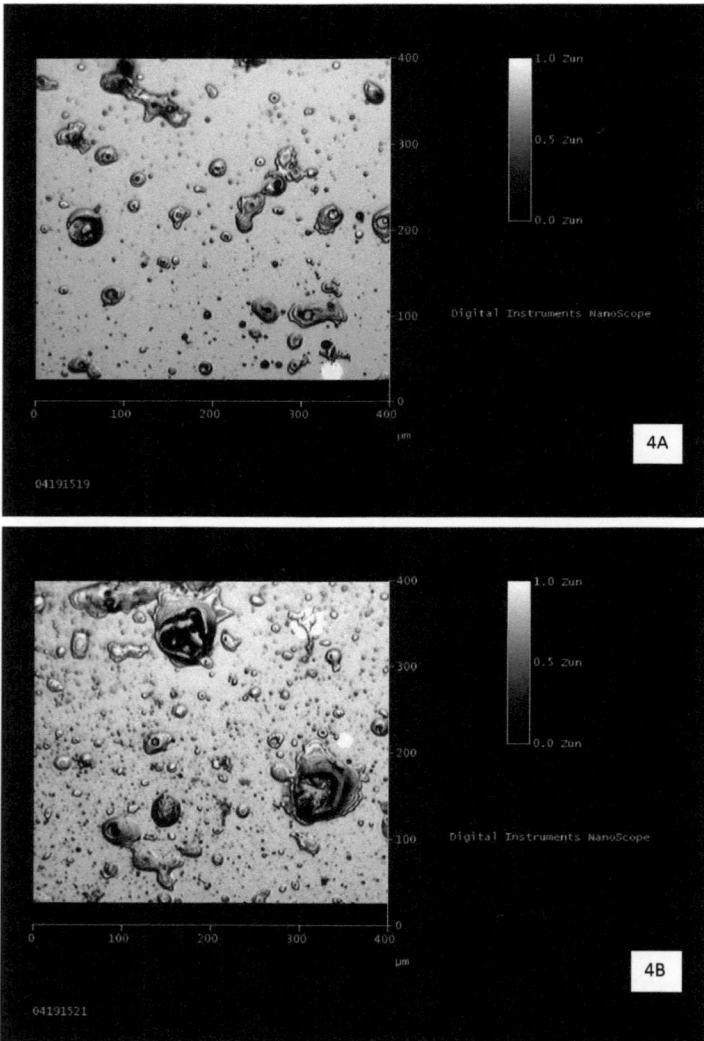

Figure 4 - Microcontainers formed by electropolymerisation of pyrrole on bare gold using condition 2: Beta-naphthalenesulfonic acid concentration = 0.4M

activities and acidity of beta-napthalenesulfonic acid in improving conductivity[1]. Hydrogen bubbles formation could be seen by eye around counter electrode due to reduction of proton[1]. However, electrolyte-pyrrole solution was too opaque to further observe if these later formed templates on gold surface during positive scans. Similar curve was obtained with 0.4M beta-napthalenesulfonic acid (not shown).

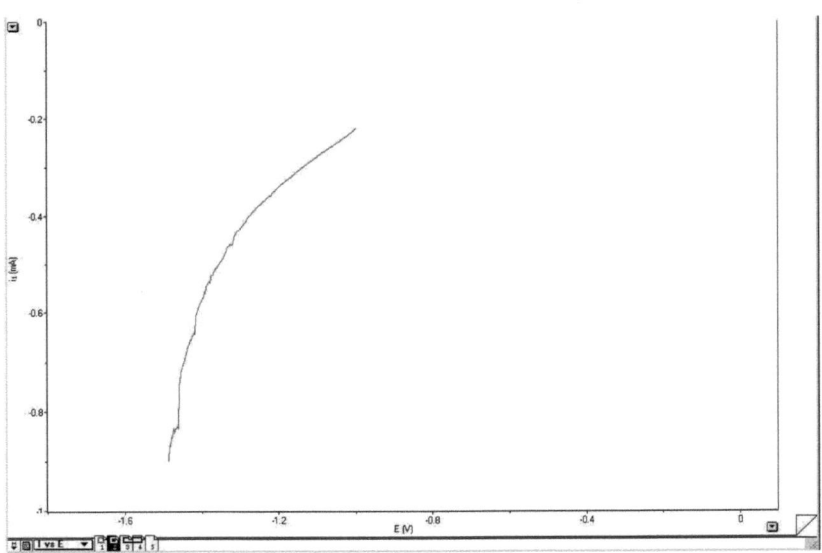

Figure 5 – Current i(mA) vs Potential(V) vs Ag/AgCl diagram for first cyclic voltametry scan of -1.0V to -1.7V to generate hydrogen gas template for bare gold using 0.2M beta-naphthalenesulfonic acid using 0.2M beta-napthalenesulfonic acid.

Evidence of pyrrole polymerization on gold electrode could be seen from increased conductivity during two scans of positive potential (Figure 6). In the first cyclic voltametry scan, as voltage was increased linearly from 0.5 to 0.9V current increased hyperbolically. This was followed by increased current for same amount of potential applied when potential was decreased linearly from 0.9V to 0.5V. Pyrrole were oxidized to polypyrrole and become conductive. As soon as polypyrrole started forming structures on gold surface, polypyrrole added cross sectional surface area for conducting electricity and hence, increased conductivity[5]. The second scan showed similar trend with more current per voltage

19

applied than the first, implying more polypyrrole deposition that increased conductivity. With increased beta-napthalenesulfonic acid concentration (0.4M), the curve showed similar trend (Figure 7) but was off-scale due to unexpected more current. Interestingly, in the second scan, the amount of current seemed decreased for the same applied potential. This implied that conductivity slightly decreased. This could be attributed to loose welding connection between counter electrode's wire and coil. Eventhough both coil and wire are immersed, loose connection would reduce current supplied by counter electrode due to reduced cross sectional surface area from coil-wire to wire[54].

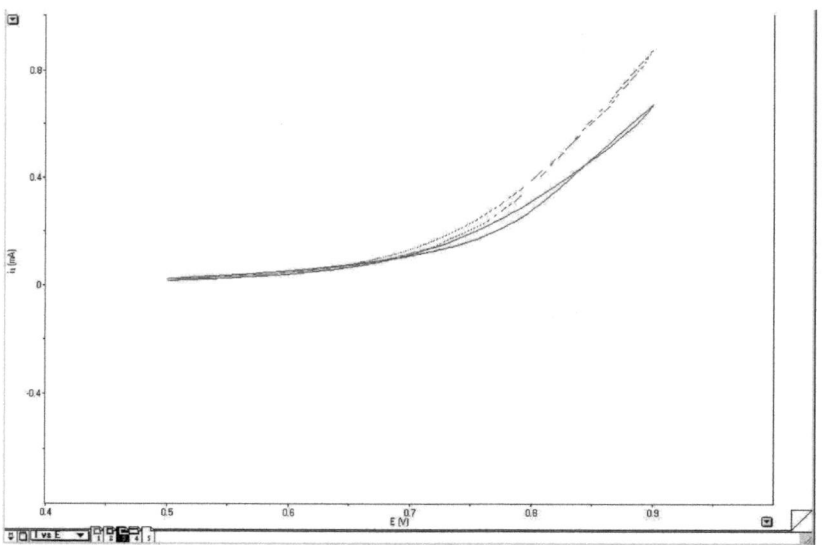

Figure 6 - Current i(mA) vs Potential(V) vs Ag/AgCl diagram for two cyclic voltametry scan of 0.5-0.9V to start polymerizing pyrrole template for bare gold using 0.2M beta-naphthalenesulfonic acid. Dashed line was the second scan.

5.3 – Surface hydrophobicity

Gold was found to have similar electrolyte-pyrrole contact angle to thiol coated gold of ($\theta_{solution} \sim 70°$) (Table1). This implied that gold had carbonaceous contamination and this should have been cleaned by UV. This also affected result in previous section. Furthermore, goniometer was broken so contact angle measurement was approximated by eye. Highly oriented pyrolitic graphite was found to be hydrophobic with high contact angle ($\theta_{solution} \sim 90°$)(Table 1). n-decane contact angle was also observed to be fairly low ($\theta_{n-decane} < 30°$) on highly oriented pyrolitic graphite and thiol-coated gold. If graphite was more hydrophobic, then $\theta_{n-decane}$ on graphite < $\theta_{n-decane}$ on thiol-coated gold. Therefore, microcontainers formed on graphite would have more hemispherical structure than that on thiol-coated gold.

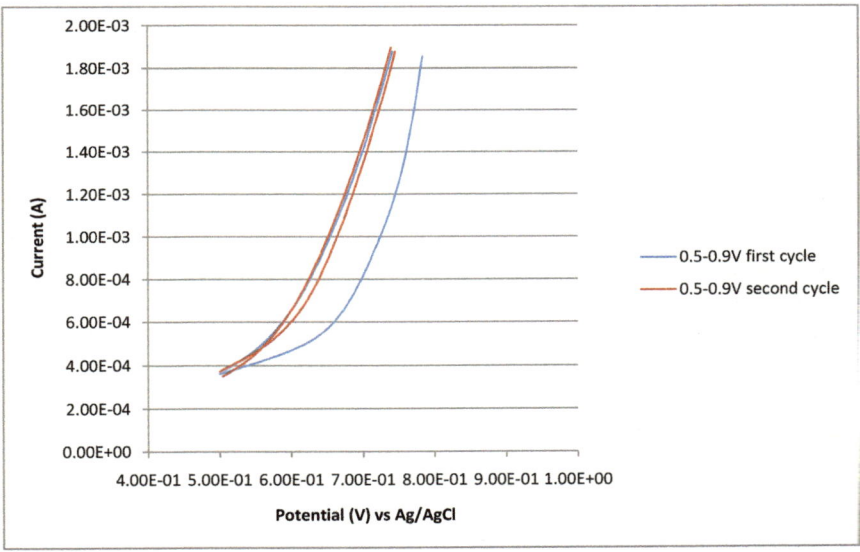

Figure 7 - Current i(mA) vs Potential(V) vs Ag/AgCl diagram for two cyclic voltametry scan of 0.5-0.9V to start polymerizing pyrrole for bare gold using 0.4M beta-naphthalenesulfonic acid.

5.4– Electropolymerisation of pyrrole around n-decane droplet template on highly oriented pyrolitic graphite

Unfortunately, there was no interesting microcontainer on highly oriented pyrolitic graphite. Both working and counter electrode was thickly coated with black polypyrrole film[4] after two positive potential scans. Also, no interesting graph was produced due to overpotential notification from the potentiostat. This supported the study that related larger size of

Table 1. Static sessile contact angle measurement using goniometer

Solution and surface	Run	Static right contact angle (degree)	Static left contact angle (degree)
0.2M beta naphtalenesulfonic acid	1	63.9	66.1
	2	76	80.5
and 0.25M pyrrole on bare gold	3	73.8	69.9
mean value		71.2	72.2
0.2M beta naphtalene sulfonic acid	1	false tangent: 168.8	60.1
	2	false tangent: 39.6	75.7
and 0.25M pyrrole on thiol-coated gold	3	false tangent: 36.4	74.9
mean value		false value: 81.6	70.2
0.2M beta naphtalene sulfonic acid and 0.25M pyrrole on highly oriented pyrolitic graphite		picture cannot be taken due to	
		broken goniometer, approximately	
		90 degree by eye	
n-decane on graphite		picture cannot be taken due to	
		broken goniometer,	
		less than 30 degree by eye	
n-decane on thiol-coated gold		picture cannot be taken due to	
		broken goniometer,	
		less than 30 degree by eye	

working electrode than counter electrode to polarization[1]. Furthermore, from contact angle measurement of n-decane on it, any successful structures formed would be quite hemispherical. This meant structure observation would be possible using available microscope, but more advanced imaging would be required to examine these in detail.

5.5– Electropolymerisation of pyrrole around n-decane droplet template on thiol-coated gold

Interesting island structures were produced on thiol-coated gold (Figure 8 and 9). Both applied cyclic voltametry conditions did not seem to show much difference on structures featured. According contact angle measurement, the predicted structure was hemispherical capsule and the common size of this varied in one study from 5-10 micrometers[55]. Irregular common structures with this size could be observed, however advanced equipment was needed to see if these were hemispherical capsules. Microscope with 1000x magnification was used to observe some structures. Colour showed interference[4], and

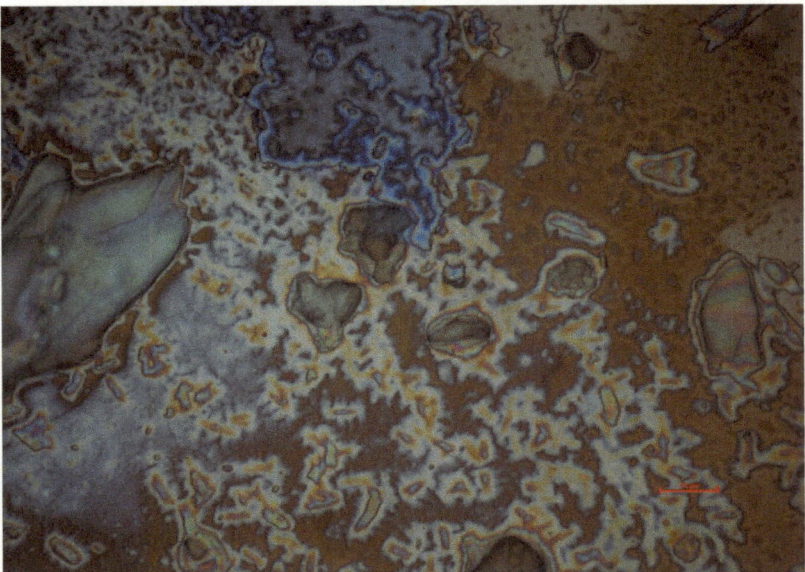

Figure 8 - Microcontainers formed by electropolymerisation of pyrrole on thiol-coated gold using condition 3: Cyclic voltametry scan of 0.5V-0.9V for 2 cycles. Scale bar = 10 micrometer

rainbow colour seemed to show to immiscibility of n-decane in electrolyte-pyrrole solution. If this was true, this could be used to trace decane. However, polymerized structure could also have hidden this colour. Gold instability seemed to cause many irregular structures on the surface itself and the structures formed.

Figure 10 showed positive linear trend of current versus potential during two cyclic voltametry scan of 0.5-0.9V. Surprisingly, both cycles gave the same result. No change in measured current tempted to imply constant conductivity. In other words, it could be implied that there was no addition in surface area to conduct electricity, because there was no polymerization. However, some significant structures were produced on Figure 8. Interestingly, applied positive potential gave negative measured current. A different condition (two cyclic voltametry scan of 0.4-0.6V) produced similar structures. However, current versus potential curve showed overpotential (not shown), as opposed to negative current. After this experiment, counter electrode incidentally disconnected at wire-coil welding point. This showed fairly loose connection at the counter electrode during

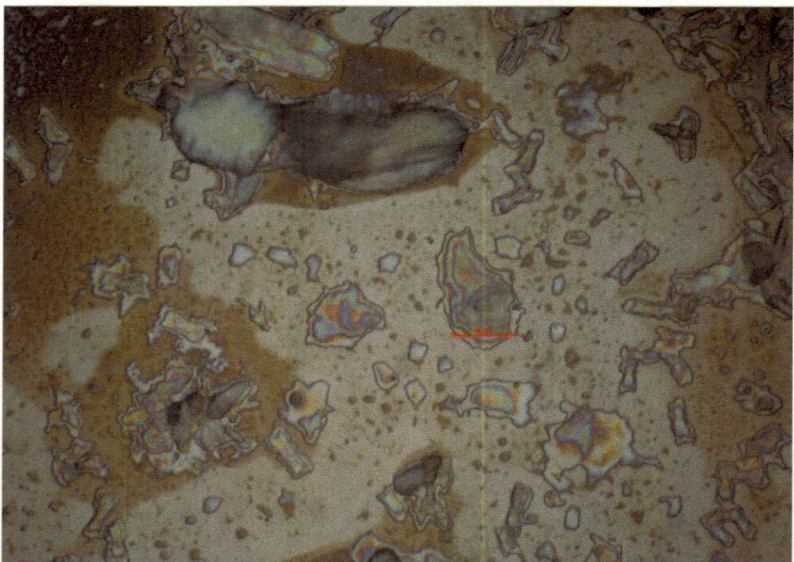

Figure 9 - Microcontainers formed by electropolymerisation of pyrrole on thiol-coated gold using condition 4: Cyclic voltametry scan of 0.4V-0.6V for 2 cycles. Scale bar = 10 micrometer

experiment, which could explain inaccuracy in current measurement. Nevertheless, polymerization was believed to occur by observing irregular structures formed.

One hypothetical difference between the outcomes of different conditions was with higher applied potential, the suspected n-decane template seemed to be more sealed by pyrrole. This could be seen from rather dark colour on suspected structures in Figure 8, implying pyrrole polymerization/oxidation. Whereas, structures in Figure 9 seemed to still contain traces of n-decane, which is inferred from its rainbow color. This could mean collapsed structure or n-decane not sealed by polypyrrole. Another unknown variable was whether anionic surfactant electrolyte coated n-decane droplet with negative charges, or instead, started to polymerise as film on the working electrode. Poly-beta-naphthalenesulfonic acid film emitted blue colour, which existed in Figure 8[6]. This interpretation, however, was based on a weak assumption because interference of colour was not fully understood and could be misleading.

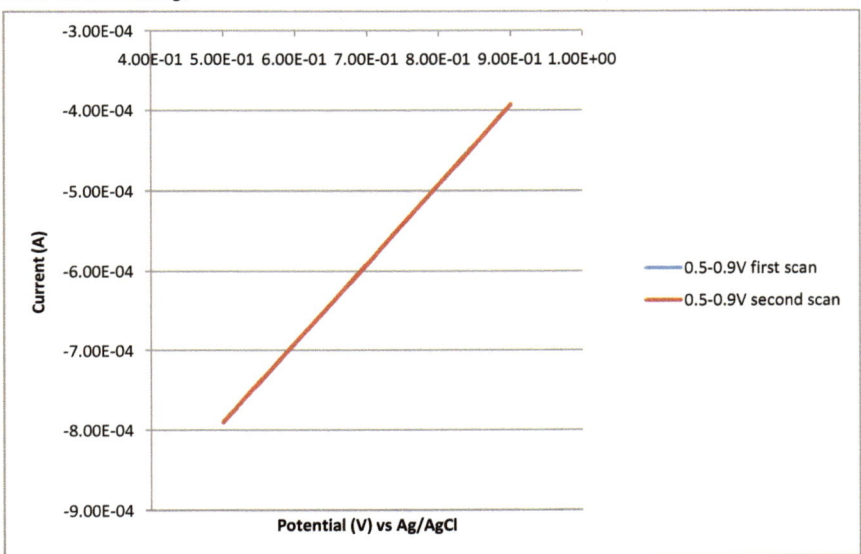

Figure 10 - Current i(mA) vs Potential(V) vs Ag/AgCl diagram for two cyclic voltametry scan of 0.5-0.9V to start polymerizing pyrrole for thiol-coated gold using 0.2M beta-naphthalenesulfonic acid.

6 –Conclusion

Simple electrochemical polymerisation of pyrrole produced observable microcontainers on sputtered gold and thiol-coated gold. Thick polypyrrole film was covering highly oriented pyrolitic graphite, instead.

Containers as large as hydrogen bubble size was observed with 0.4M electrolyte concentration on gold but doubted to be produced by polymerizing hydrogen bubble as a template. Instead, 10 micrometer size containers seemed to be common containers produced for 0.2M and 0.4M electrolyte concentration, possibly with other templating model such as pyrrole droplet[48]. Gold instability was observed, and suspected to contribute to microcontainers resulted, such as in behaving as non-homogenous surface. Contact angle measurement implied carbonaceous contamination on gold surface before polymerisation, which could surface properties and result too.

Highly oriented pyrolitic graphite size, which was larger than counter electrode, caused polarization on this working electrode. This meant it underwent chemical reaction in faster rate, which lead to overoxidation of pyrrole. If size had been reduced, expected structures would have been rather flat capsules due to high hydrophobicity of graphite implied in contact angle measurement. This would have been difficult to observe with available equipments.

Irregular polymerized capsules were observed with n-decane template formation on thiol-coated gold surface with two different cyclic voltametric scan range. Both did not show significant This is in contradiction with contact angle measurement implication of hemispherical capsules. Again, gold instability could have caused unexpected microcontainer shape. Another weak hypothethical reasoning considered based on interference of colour was n-decane trace on rating successful template polymerization and possible polymerization of electrolyte on surface due to blue colour observed.

Succesful polymerisation of pyrrole was measured from increased conductivity from current versus voltage curve, which appeared in later to be inaccurate due to counter electrode wire-coil welding looseness. While simple electrochemical polymerization were feasible to do using basic materials and equipments, these also caused some variables were not satisfactorily controlled. They were distance between working, reference, and electrode, graphite electrode size relative to counter electrode, instability of gold electrode, immersion of copper wire attached to working electrode in electrolyte-pyrrole aqueous solution, cleanliness of Ag/AgCl reference electrode to accurately measure applied potential[4], and counter electrode wire-coil connection which seemed to wear-off during polypyrrole cleaning by flaming and hence affected accuracy in measuring current[5].

Therefore, eventhough this work has not been conducted in well controlled manner, some results have shown that microcontainers were successfully fabricated as aimed. Obviously, better equipments and material would have eliminated inaccuracy and made results more observable, such as stable gold, more advanced electrochemical system and microscope, spectroscopy, and imaging devices.

7 – Reference

[1] C. Li, H. Bai and G. Shi. Conducting polymer nanomaterials: electrosynthesis and applications. Chem. Soc. Rev., 2009, 38, 2397–2409

[2] V. Bajpai, P. He, and L. Dai, Conducting-Polymer Microcontainers: Controlled Syntheses and Potential Applications. Adv. Func. Mater. 2004, 14, No.2, February

[3] L. Qu and G. Shi. Electrochemical synthesis of novel polypyrrole microstructures. CHEM. COMMUN., 2003, 206–207

[4] X.H. Zhang. Discussion during Research Project. 2010.

[5] S.P. Best. Discussion during Research Project. 2010.

[6] L. Eh et. al. Electrochemical polymerization of beta-naphthalene sulfonic acid. JOURNAL of APPLIED POLYMER SCIENCE 92 (3): 1939-1944 MAY 5 2004

[8] L. Dai, Radiat. Phys Chem. 2001, 62, 55.

[9] N. Sakmeche, S. Aeiyach, J. J. Aaron, M. Jouini, J. C. Lacroix and P. C. Lacaze, Langmuir, 1999, 15, 2566.

[11] R. J. Jackman, G. M. Whitesides, CHEMTECH 1999, 18.

[13] L. Dai, P. Soundarrajan, T. Kim, Pure Appl. Chem. 2002, 74, 1753.

[16] R. A. Erb, J. Phys. Chem. 1965, 69, 1307.

[19] E. Thelen. J. Phys. Chem. , 1967, 71, 4578

[25] R.S. ROBINSON, et. al. MORPHOLOGY AND ELECTROCHEMICAL EFFECTS OF DEFECTS ON HIGHLY ORIENTED PYROLYTIC-GRAPHITE. JOURNAL OF THE ELECTROCHEMICAL SOCIETY 138 (8): 2412-2418 AUG 1991

[26] A. Agnihotri and C.A. Siedlecki. AFM Investigation of Molecular Level Events at Biomaterial-Biological Interface. Bioengineering, Proceedings of the Northeast Conference, p 125-126, 2002.

[30] J. Cognard. Adhesion to Gold: A Review. Gold Bull., 1984, 17, 4.

[31] L. Rothschild. The Polarization of a Calomel Electrode. The Royal Society, 1938, CXXV, B., 283.

[32] J.R. Vig. UV/Ozone Cleaning of Surfaces. US ARMY LABORATORY COMMAND. May 1986.

[33] F. Hui et al. Electrochemical fabrication of nanoporous polypyrrole film on HOPG using nanobubbles as templates. Electrochemistry Communications 11 (2009) 639–642.

[45] T. Hatano, M. Takeuchi, A. Ikeda and S. Shinkai, Org. Lett., 2003, 5, 1395–1398.

[46] G. W. Lu, C. Li and G. Q. Shi, Polymer, 2006, 47, 1778–1784.

[47] S. Gupta. Template-free synthesis of conducting-polymer polypyrrole micro/nanostructures using electrochemistry. APPLIED PHYSICS LETTERS 88, 063108 2006

[48] Y. Gao et al. Electrosynthesis of small polypyrrole microcontainers. Journal of Electroanalytical Chemistry 597 (2006) 13–18

[49] J.T. Kim et al. The microcontainer shape in electropolymerization on bubbles. Appl. Phys. Lett. 94, 034103 (2009)

[54] E. Smela. Microfabrication of PPy microactuators and other conjugated polymer devices. 1999 J. Micromech. Microeng. 9, 1

[55] M. Mazur. Preparation of Surface-Supported Polypyrrole Capsules Using a Solidified Droplets Template Approach. J. Phys. Chem. B, Vol. 113, No. 3, 2009

[56] M. Mazur. Polypyrrole Containers Grown on Oil Microdroplets: Encapsulation of Fluorescent Dyes. Langmuir 2008, 24, 10414-10420

[57] D.G. Shchukin, K. Ko¨hler, and H. Mo¨hwald. Microcontainers with Electrochemically Reversible Permeability. J. AM. CHEM. SOC. 2006, 128, 4560-4561

[58] T. Smith. The hydrophilic nature of a clean gold surface. Journal of Colloid and Interface Science, Volume 75, Issue 1, May 1980, Pages 51-55

[59] C.-K. Kang Y.-S. Lee. The surface modification of stainless steel and the correlation between the surface properties and protein adsorption. J Mater Sci: Mater Med (2007) 18:1389–1398

[60] A. Sarapuu, et. al. Electrochemical reduction of oxygen on nanostructured gold electrodes. Journal of Electroanalytical Chemistry, Volume 612, Issue 1, 1 January 2008, Pages 78-86

[61] L. Jing and W. Erkang. Scanning tunneling microscopy (STM) and lateral force microscopy (LFM) investigation of the structure and character of conductive polypyrrole film. Synthetic Metals 1994 66, 1, 67-74